Leslie Trietlin began his working life as an engineer after gaining a degree in Technology. He then moved into the world of sales and spent the next 25 years as a self-employed selling agent. He lived for five years in Spain, where he was the vice-president of the British Society of the Costa del Sol, and for seven years in the Dordogne region of France. He currently lives in a village in the Cotswolds.

MATHS IN EVERYDAY LIFE

Leslie Trietlin

Book Guild Publishing
Sussex, England

First published in Great Britain in 2005 by
The Book Guild Ltd
25 High Street
Lewes, East Sussex
BN7 2LU

Typesetting in Times by
Keyboard Services, Luton, Bedfordshire

Printed in Great Britain by
Optigraph Ltd, Crowborough, East Sussex

A catalogue record for this book is available from
The British Library

ISBN 1 85776 906 6

CONTENTS

PREFACE

The purpose of this book is to explain, and hopefully teach, the basics of mathematics, in terms of the relationship between percentages, decimals, and fractions, in a simplified and more understandable way.

In the modern world, millions of calculations have to be made hourly, mainly by computer, the use of which is becoming ever more important for anybody who works in a bank, office or almost any type of industry. The fact that computers do the actual calculations, does not obviate its user needing to know how to feed in the correct information. Since decimalisation it has become even more important to have an understanding of decimals and percentages, since the two are inextricably linked in all monetary transactions. It follows also that an understanding of the relationship between these two and fractions is extremely useful.

To give an example, it was stated by the government during the year 2000 that all students wishing to take teacher training must be able to answer the following question: 'what percentage is thirty six of sixty?' I put this question to several of my friends, all of whom were graduates, some teachers, and not one could answer correctly; the best was, 'a bit more than half'. The answer of course is sixty per cent.

This exercise convinced me that basic understanding of this type of calculation is not very well implanted in most people's minds, and that a better way of teaching or explaining the principles involved is necessary. After thinking about it for some time, I came to the conclusion that the way we are taught at school does not help to give the student a clear and lasting understanding of the underlying principles. Most people do not seem to understand that there is an important relationship between fractions, decimals and percentages. The tendency is to teach them as separate topics, without stressing the importance of knowing how they interrelate.

In today's world we all use calculators, but how many people know after they have used a calculator whether the answer it shows is the correct one? I am not suggesting that it has malfunctioned, but that the user may have hit the wrong button, or fed in the wrong information, or just not understood the correct way to program the calculator in the first place. So, without an understanding of what you are doing, it is impossible to know whether the answer given is right or wrong.

It is also important to realise that a calculator or a computer can produce calculations in percentages, decimals, or fractions, so long as the inputs are converted into the language that the machine can understand. In this respect, it is crucial to understand the transition from one to the other.

One of the problems about learning mathematics is that young people are inundated at school with all types of calculations in large numbers, so that for many students the whole thing becomes a complete jumble of figures impossible for the mind to accept. Consequently, maths becomes a hated subject, attention is not paid to learning it, and many people go through life never understanding even the basic concepts.

1

TERMINOLOGY

FRACTION

A fraction is a numerical quantity that is not a whole number. A fraction always has to be less than one, or less than a quantity of any size (that is being treated as a whole).

DECIMAL

Of tenths, or ten. Counting in tens. Reckoning or proceeding in tens.

DECIMAL SYSTEM

Each denomination is ten times the value of the one before it.

DECIMAL POINT

The dot placed after the unit figure.

PER CENT

Meaning 'of a hundred', it comes from the Latin word *centum* meaning century. Roman legions were made up of centuries (one hundred men), the person commanding a century being called a centurion.

PERCENTAGE

Proportion of a hundred. The symbol used for representing percentage is %, 100% being used as the basis for all percentage calculations.

2

THE DENARY SYSTEM

The Denary system is our basic method of counting. Denary means ten, and is represented by the following ten numbers.

1 ONE	2 TWO	3 THREE	4 FOUR
5 FIVE	6 SIX	7 SEVEN	8 EIGHT
9 NINE	0 NOUGHT		

These ten numbers make up all the combinations of numbers that we use in our counting systems, whatever these systems may be. All these numbers are used for calculating in fractions, decimals and percentages.

Two of these numbers are very special because they can represent more than their initial meaning. For instance, one obviously means one – one thing, one item, etc. And nought means nothing; but one (1) can also represent the whole, whatever the whole may be, whether it is a person or a very large number. I will discuss this in more detail later. Nought, as I have stated, means nothing by itself, but add this to

13

any other number and it multiplies that number by ten; so one plus a nought equals ten, or ten times one (10×1). Add another nought and we have 100, or ten times ten (10×10). So with the addition of every nought any number is multiplied by ten; this is a very important concept.

This multiplying of a number by ten only applies if the nought is added to the right of that number; if the nought is added to the left of the number, it has no value unless a decimal point is placed between the two. For example, 010 means ten, and 0010 also means ten, but if a decimal point is added, 0.1 means one tenth, and 0.01 means one hundredth. So the decimal point makes all figures to the right of the decimal point negative, that is, less than one. Thus any number to the right of a decimal point is less than one (1), and is therefore a fraction.

For instance we have said that .1 means one tenth: therefore it is the fraction one over ten, i.e. 1/10. If a further nought is added between the decimal point and the one (1), this is multiplied by ten, but in a negative way. So point nought one (.01) is now ten times ten, which equals one hundred (100); but because it is negative it now means one hundredth, the fraction one over a hundred, i.e. 1/100.

So the rule is that every nought added to the right of the decimal point, between the decimal point and whatever number is there, makes that number ten times less, so there is a direct correlation between decimals and fractions.

.1 = 1/10
.01 = 1/100

.001 = 1/1000
.0001 = 1/10000

You can see by the above that every nought added to the decimal side also adds a nought to the bottom side of the fraction (the denominator).

Looking at the positive side again, as was seen previously, a nought added to the right of a positive number multiplies that number by ten, and each succeeding nought multiplies the number again by ten, so that we have:

1 + 0 = 10 = 1 × 10 ten
10 + 0 = 100 = 10 × 10 one hundred
100 + 0 = 1000 = 100 × 10 one thousand

and so on.

Now let us look at a large number, say, 00000100000; this looks very daunting, but let us analyse it. We know that this number must be positive because there is no decimal point. But we also know that all the noughts to the left of the one (1) have no value, so that we can rewrite this number as 100000, which is one hundred thousand.

If we now take the number again and insert a decimal point between the third and fourth noughts from the left, i.e. 000.00100000, we know that all the noughts to the left of the decimal point have no value, because there is no positive number; and that all the noughts to the right of the one (1) also have no value because the one (1) in this instance is negative; and that only the noughts to the left of the one (1), between the one (1) and the decimal point, have any value. Each nought effectively makes the value ten times less, so we can rewrite this number as .001, which is equal to one thousandth.

If we take the number again and insert a decimal point between the first and second noughts to the right of the one, (1) i.e. 0000010.0000, all the noughts to the left of the one (1) have no value, and all the noughts to the right of the decimal point have no value; so we can rewrite this number as 10 (ten).

It is possible to give the above string of eleven numbers twelve different values depending upon where the decimal point is placed.

So now try the following exercises.

Exercise 1

Rewrite and simplify the following numbers, quoting them in figures, words and fractions if applicable:

    a. 000.00100000
    b. 0000010.000
    c. 000001000.00
    d. 00000.100000
    e. 0000010000.
    f. 0000.0100000
    g. 00000100.000

Before we can go further it is important to look at the other nine numbers in the denary system, and their relationship to the following concepts. Their values are very obvious in the positive. We all have learned the value difference between ten (10), twenty

(20), thirty (30), or one hundred (100), two hundred (200), three hundred (300), and so on. But they are a different concept in fractions and negatives.

For instance, one half of ten (10) in fractions is one over two ($^1/_2$), but in decimals it is point five (.5). To prove this we need long division. We have to divide one by two ($^1/_2$) the sum in long division form is 2\1\; two will not divide into one so the answer is 2/1/0; the nought is because it is impossible to divide two into one. We now have to add a nought to the one to make it divisible, but it now becomes a negative sum, so we must put a decimal point after the nought in the answer, so now the sum is 2\10\0.5 so the answer is point five (0.5) or (.5). Both are the same as we know.

Any fraction is a negative number because it is only a part of the whole number, irrespective of the size of that number, and in decimal terms it will always be on the right side (the negative side) of the decimal point. All the numbers in the denary system are also used in the negative, and have their own values accordingly.

The decimal value of any fraction can be found by dividing one (1) by the denominator 'the lower number of the fraction', for instance, one (1) divided by two (2) = $^1/_2$ which is .5 as a decimal. The simplest way to find the decimal of a fraction is to use a calculator. The sum is as follows, $1 \div 2 = .5$, $1 \div 4 = .25$. When the fraction is larger, for instance 7/8, first divide one by eight, then multiply by seven, $1 \div 8 \times 7 = .875$.

It is important to learn this concept thoroughly, because when using a calculator it is impossible to use fractions as fractions. Calculators cannot understand fractions; they only understand decimal equivalents of fractions, and when working in percentages, which we are going to look at next, all the calculations will involve using decimal equivalents of fractions. To give an example, a calculator could not process the number that is two ninths of one hundred and eighty (2/9 of 180), so one could simply divide one hundred and eighty by nine (180 ÷ 9) which equals twenty 20, and multiply the resulting answer (20) by 2, which equals 40.

Exercise 2

Using a calculator, change the following fractions into decimal equivalents:

a. One eighth (1/8)
b. One third (1/3)
c. One ninth (1/9)
d. One fifth (1/5)
e. One seventh (1/7)
f. One fifteenth (1/15)
g. One twentieth (1/20)
h. One thirty fifth (1/35)
j. One fortieth (1/40)
k. One fiftieth (1/50)
l. Five eighths (5/8)
m. Five sixths (5/6)
n. Four sevenths (4/7)
p. Seven ninths (7/9)
r. Three eighths (3/8).

3

PERCENTAGES

Per cent means of a hundred. It is the method by which increases or decreases are calculated in numerical terms in a variety of situations; for instance, in all monetary calculations, banking, income tax, inflation or deflation, economic growth, growth or decline in the national product, etc. Percentage can also be used to calculate the performance or efficiency of anything that can be quantified; for instance, machinery or even the performance of people.

The reason why percentage calculation is such a useful device is that anything can be considered as a whole by giving it the value of one hundred and calculating the increase or decrease as a percentage of the whole against the base of one hundred, the one hundred being 100%.

Looking now at figures, we can see that an increase of 1% on the sum being calculated is also equal to an increase of 1/100, one hundredth. An increase of 10% is therefore equal to 1/10, one tenth, and an increase of 100% is equal to double the original figure. Not all things go up in value, so it is also possible to have decreases, and these also can be calculated in percentages. So if something lost 10% of its value, it would be worth 1/10 less than the previous value, so that £100 reduced by 10% would be worth only £90.

In the modern world, more and more of us are becoming involved daily in money transactions of all kinds, many of these involving percentage calculations, so it is important that everybody has a knowledge of how percentage calculations work. The man in the street may consider it unimportant to understand percentages, but if you have a mortgage, or borrow money, or deposit money in a bank, it is very important to know the rate of interest that you are either paying or receiving.

For instance, credit cards are now a major item in everyday life and a typical example of how people can be lured into using cards that charge excessive rates of interest. Some cards have a rate of 29% (if you read the small print), which is truly an astronomical rate of interest. So let us look at some calculations. We will keep it simple to begin with, and use only the figures one and nought (1,0).

100% is the base figure for all percentage calculations and, as stated before, can represent any quantity, or any whole. Suppose we consider a bank account of £100 (a nice round figure for the 100%). If this amount was on deposit at an annual rate of interest of 10%, it would accumulate £10 interest in one year. If the amount on deposit was £1,000 at the same rate, it would obviously accumulate 10 times the amount of the £100 figure (with interest of £10), that is, £100 interest. So at the end of the year the capital would have grown to £1,100, (£1,000 + £100). If this amount was left on deposit for a further year at the same rate, it would become £1,210. This is calculated simply as £100 interest on the £1,000 plus £10 interest on the £100, making an increase of £110.

There is also a simpler way of calculating a 10% increase on any figure. We know from before that a nought at the end of a row of figures multiplies that number by 10. So all that needs to be done is to eliminate the final nought, and what is left becomes the 10% increase. Thus from our figure of £1,100, minus the nought we get £110, and this is the interest.

Consider now the opposite of deposits, that is, borrowing money. If £1,000 is borrowed at a rate of 10%, at the end of the year the borrower will have had £100 added to the amount owing, making the sum owed £1,100. If the capital is not repaid on this amount, it will continue to have interest added until it is repaid.

At this point we should consider the difference between two types of interest calculation

1. Simple interest
2. Compound interest.

The dictionary states that with simple interest, 'interest is paid on the principal alone'. With compound interest, 'interest is calculated on both the principal and its accrued interest'.

In the case of deposits, compound interest is fine: the longer we leave our money in the bank, the more it grows, because the accrued interest is accruing more interest (i.e. more money). But in the case of borrowing money from a bank or a moneylender, the opposite occurs, because the lender starts to add interest to the capital loaned, as soon as the loan is given to the borrower. Therefore, the amount to pay back is forever on the increase until the whole of the loan plus the accrued interest is paid back in full.

Interest paid on bank deposits is simple interest, which is calculated on a one year term. So in the case of an interest rate of, say, 4%, £4 will be added at the end of each year to every £100 that is on deposit. Similarly, if a sum is borrowed, then the interest will be added at the end of each year to the amount still outstanding. Unfortunately, borrowing rates are always considerably higher than those available on deposit.

When using a calculator the easiest way to add the interest to a capital amount for one year is: first enter the amount into the calculator, then press the plus sign, then press the numbers for the interest required, then press the percent sign (%). For instance, for £280 @ 3.4%, the sum is 280 + 3.4%. The answer will be shown completed.

Exercise 3

Calculate the following simple interest for one year:

    a. £634 @ 4.5%
    b. £372 @ 5.75%
    c. £1,214 @ 6.25%
    d. £3,421 @ 11%
    e. £2,731.43 @ 5.1%

Beware of borrowing money at a rate that is calculated on a shorter period than the standard annual rate. Even banks have been known to lend money on a monthly rate, the monthly rate being one twelfth (1/12) of the yearly rate. This may seem to be exactly the same, but because of the compounding effect, it is not. Consider an annual rate of 12%. At a monthly rate this would be divided by twelve, so would become 1% a month. But since this would be added to the amount outstanding monthly, it would become 12% plus the accrued monthly interest, which would make it 12.682% over the year. Borrowing over several years at a monthly rate would increase the amount considerably. Loan sharks, the term used for backstreet lenders, often have weekly rates of interest which can be very expensive indeed.

Let me give you an example of the difference between annual and monthly rates. Suppose that you wanted to borrow £1,000 and you were quoted an annual borrowing rate of 20%. At the end of one year you would obviously owe £1,200. But suppose you had difficulty in borrowing the money, and a friendly loan shark offered to lend you the £1,000 at the rate of 5% interest monthly. This looks very attractive. 5% is much better than 20%, isn't it? Well, no it is not. Here's how it works out.

In one month you will owe £1.050; in two months, £1,102.50; in three months, £1,157; in six months, £1,340; in nine months, £1,551; and in twelve months, £1,795, and so it goes on until it becomes very difficult to pay back. This type of loan is only worth considering if it can be paid back very quickly.

4

FURTHER USES OF PERCENTAGES

Percentage calculations can be used in many other ways than just monetary calculations. Calculating efficiency is one of the main uses. Any input has an output, and the more output that is achieved for a given input, the greater is the efficiency. For instance, cars run on various fuels, and it is commonly accepted that diesel cars are more efficient than petrol cars; in other words, diesel cars go more miles on a gallon of fuel than do petrol cars. It is also accepted that modern cars are far more efficient than older cars, because the engines are more highly advanced.

These days almost everything is being calculated in terms of percentage efficiency: industry, railways, the health service, government, and of course pass rates in schools.

While we are on this subject let us look at a phrase in common usage: 'he gave 110% (or more) effort,' is a popular but illogical use of percentages. Nobody can give more than 100% effort, because that is all that any of us have within our body. If a person is performing at a greater rate than previously, it only means that they were not performing at their full potential before. When an athlete improves his performance, it only means that he is producing a greater percentage output to his input. This may be achieved either by training harder, more appropriate food intake, or taking performance enhancing drugs.

Percentages can be calculated either by addition or multiplication. So far we have only dealt with addition; for instance, adding 10% to 100 = 100 + 10 = 110. But calculating a complex percentage is much more difficult, for instance, 6.3%, where the number is large, say, 1,489. For most people this involves the use of a calculator. If it has a percentage button the sum is easy, as follows: first enter the sum 1,489, then press the plus sign, then enter 6.3, then press the % sign. So, 1489 + 6.3% sum completed = £1,582.80

Percentage calculations can also be done by multiplication. To do this we have to remember that any quantity can be considered as 100%, and this can be substituted by a figure 1 in decimal terms. Therefore, the percentage figure to be calculated becomes a negative decimal point place. So in the case of increasing by 10%, this becomes decimal .1 and the whole sum becomes 1.1, then all that needs to be done to make the calculation is to multiply the capital figure by 1.1, therefore, using a calculator we have 100 × 1.1 = 110, the first 1 repeats the capital figure and the .1 adds on the interest. The percentage figure can be anything up to .99999 etc which is 99.999% etc. A percentage figure less than 10% would be represented by a figure to the right of .0; for instance, 5% would be .05, 3.5% would be .035, and so on. So to find the interest on a capital figure of, say, £14,376.20 at 3.5%, the sum is 14376.20 × 1.035 = 14879.367.

At this point it would be a good idea to clarify the decimalisation again. One of the most difficult concepts to understand is the relationship between decimals, percentages and fractions. Increasing anything by 100% does not mean multiplying by 100; it merely means doubling, or multiplying by two. This is because the 100 means the whole or one of anything. Anything less than 100 can be deemed a fraction or a decimal, so that 50, can be thought of as either $\frac{1}{2}$ or as .5; both mean the same.

Suppose we had a sum where we had to increase a figure by 150%. Let's take the above figure of 14,376.20; the calculation would be 14376.2 × 2.5 = 35940.5; the 2 has doubled the capital figure, making an increase of 100%, and the .5 has added 50% interest, making a total of 150% increase.

The rule on the positive side of the calculation (to the left of the decimal point) is that any number above 1 is multiplying the capital amount, so 2 means doubling. 3 means trebling, 4 means quadrupling and so on.

To the right of the decimal point all figures are less than 100%, or less than 1, or a fraction of 1; so .1 = 10% =1/10, and .01 = 1% = 1/100.

Percentage loss

So far the subject of percentage loss has not been dealt with, but unfortunately percentage loss is just as frequent as percentage gain and just as important. The method of calculating a percentage loss is different from calculating a percentage gain, and it is very important that this is learned correctly or you will have some very wrong answers.

The first stage is to subtract the reduced figure from the original figure. Suppose the original figure was £1,000 and it had reduced to £800; the sum therefore is 1000 − 800 = 200.

The second stage is to divide the original amount by the amount of decrease. The sum now is 1000 ÷ 200 which gives us five (5).

The third stage is to divide 100 (because we are calculating a percentage) by the five (5). So the sum is 100 ÷ 5 = 20; thus the decrease from 1,000 to 800 is 20%.

Example	400 − 300 = 100
	400 ÷ 100 = 4
	100 ÷ 4 = 25%
Example	875 − 732 = 143
	875 ÷ 143 = 6.118
	100 ÷ 6.118 = 16.34%

Calculators and fractions

It is important to remember that all calculations nowadays are carried out on computers or calculators, but it is impossible to use fractions when using mechanical devices, computers or calculators, so all calculations in percentages, by additions, subtractions or multiplications have to be in decimals. But, even though we do not use fractions, it is important to be able to understand the relationship between the three. The following are some simple examples.

Examples:

$\frac{1}{2}$ = 5/10 = 50/100 = 50% = .5
$\frac{1}{4}$ = 2.5/10 = 25/100 = 25% =.25
$\frac{1}{8}$ = 12.5/100 = 12.5% = .125
$\frac{1}{10}$ = 10/100 = 10% = .1

The rule is first to find what relationship the fraction has to 100; it is then possible to find the relationship to a decimal if necessary. This becomes very important if you want to find a percentage rate of a figure which is less than 100. As I have shown, if you want to increase by more than 100 it is easy to add on by using the percentage key. This becomes more difficult when trying to reduce a figure by a percentage or when trying to find out how much interest has been earned on a given figure. This is where the multiplication by a negative decimal figure comes into its own.

Suppose one had a large figure which has been reduced by, say, 5%. Let's take a figure of say, 7469.53; the calculation is 7469.53 × .95 = 7096.0535. Why did we multiply by .95? Well, 100 – 5 is 95 and when turned into a negative figure, it becomes .95; hence we reduce the figure by 5%. So the rule is if a figure is multiplied by a negative figure, the result is a reduction in value of the figure that has been multiplied.

Suppose we wish to find out only the interest earned

on a capital figure by a certain percentage rate. If the original capital figure is known, it is a simple deduction of the capital plus interest minus the original capital. It is more complicated if the original capital figure is unknown, but this can be calculated if the interest rate is known. This will be dealt with later when looking at VAT calculations.

Back to our basic percentage-to-decimal calculations. What is the rule? The rule is, if the percentage rate is 100% or more, it is a positive figure and goes to the positive (that is, the left) side of the decimal point; and if the percentage rate is less than 100% (even by only a fraction of one per cent), it goes to the right of the decimal point.

If the rate is 10% or more (anything from 10 to 99), then it becomes the first figure following the decimal point. So the smallest number following the decimal point will be 1, in decimal terms .1, and the largest number 9, in decimal terms .9; if the figure is less than 10%, then it will always be .0 followed by a number, the lowest being .01, the highest being .09: and when the figure falls below 1%, we add a further 0 after the decimal point. So this now becomes .00, the lowest being .001, the highest being .009, and so on. For the purposes of normal calculations I do not think that many people will find it necessary to calculate more than three decimal places.

You will find that financial papers show international currency rates calculated to four decimal places. This is quite normal, but it should be borne in mind that, when dealing in very large amounts in currency trading, even the fourth decimal place can amount to a lot of money.

Exercise 4

Change the following percentages into decimals:

    a. 13%
    b. 42%
    c. $9\frac{1}{2}\%$
    d. $1\frac{3}{4}\%$
    e. 122%
    f. 203%
    g. $8\frac{3}{4}\%$
    h. $\frac{1}{2}\%$

The Constant Function

All calculators have a constant function built into them. This is not a particular button with the word 'constant' or the key with 'c' (which is the 'cancel' key) printed on it; it is merely the way that the calculator works. To find the constant on your calculator, switch on and enter 1.1 × 100 then press the equals sign (=); the result will be 110. Without cancelling or changing anything, press = again; the result will either be 121 or some other figure. If the result is not 121 then your calculator is not in constant mode; if the result is 121, then this is the constant mode. Now enter 100 × 1.1, press = then press = again. If this is the constant mode the result will be 121.

In the above example, the 1 repeats the 100 and the .1 adds 10% to the total, and in the constant mode each press of the = sign adds a further 10%.

The system is, the constant mode either precedes the capital amount entered or follows it; that is, either 100×1.1 or 1.1×100; (it is usually the latter).

The idea of using a constant is so that percentage increase calculations, and repeat calculations, can be performed without the need to go back to square one for every individual process.

For instance, if you want to know how much interest will be earned on a sum of money over a set amount of years at a fixed interest rate, say, for instance, £5,000 invested in a fixed-interest bond for five years at a rate of 4.5%, you merely enter the sum 1.045×5000 (or 5000×1.045, whichever way the constant is), and press the = sign five times, and the answer comes up as £6,230.91p.

Try the following exercises:

Exercise 5

    a. £513 @ $4\frac{1}{2}$% over 3 years
    b. £785 @ 5% over 10 years
    c. £1,023 @ 7.5% over 4 years
    d. £15,000 @ 3% over 5 years
    e. £6,839 @ 13.5% over 7 years
    f. £4,000 @ $11\frac{1}{4}$% over 9 years

Use of a constant in VAT calculations

Before the introduction of VAT in Britain, we had Purchase Tax. This was a tax levied by the government on most items sold; the tax varied depending upon the type of product. Purchase Tax was a once only tax, which was added to the invoice price of articles before they reached the final point of sale. Retail shops were only required to add on their mark-up to the invoice price to arrive at their selling price, and no further action was necessary by them regarding tax.

With the introduction of VAT (Value Added Tax), a tax was added to the price of the article from the manufacturing stage through all the other stages (such as wholesale) up to and including the final selling stage. It was also levied on all services, though there were some exemptions. The onus for the collection of this tax was firmly put on every stage that goods (or services) went through. So at each stage a percentage was added to the cost of an article, then a mark-up was added for profit and/or handling, and this in turn had to have the tax added. The sum needed to calculate the final selling price of any article was: invoice price, plus VAT, plus mark-up, plus VAT. Thus every stage collected more VAT. Where this involved articles with few price differences, the sums were easy to calculate; but in the case of retail shops who had hundreds of differently priced articles, it became a major operation. So let me give you an example.

When VAT was first introduced, the rate was set at 10%, which was an easy calculation. But if you are having to work out the price of each individual calculation for hundreds of different prices, and possibly at different mark-ups, then the time involved is enormous. In the early years of VAT, I met many shopkeepers who were spending most of their time calculating each individual price as above – invoice price, plus VAT, plus mark-up, plus VAT – for each article.

As you can imagine, a lot of time and money was wasted in having to make all these calculations. But there was also an easy way out – a 'multiplication constant; not quite the one in previous pages, but one that can be calculated to fit all the differing circumstances.

What is needed for this calculation is a number that can be used to multiply the invoice price by, to reach the final figure. So going back to our percentage calculations, we start with the figure 100. Assuming VAT at 10%, we add that to 100 to make 110, we then need to add on the mark-up; let us assume it to be 50% (50% of the cost without VAT is 50). We now have 110 + 50 = 160; but we also have to add the VAT on for the 50% mark-up, and 10% (50) = 5. So the sum is 100 + 10 + 50 + 5 = 165; if we now decimalise this, we get 1.65: this is the constant for any article with a 50% mark-up, so all the shopkeeper has to do is to multiply the invoice price by 1.65 to arrive at the selling price.

Should a different mark-up be required, recalculate as necessary; this also applies if the rate of VAT changes. The current rate of VAT at the time of writing is 17.5%, which of course would be much more difficult if one had to do hundreds of calculations, but is just as easy to calculate as the 10% to make a constant: following the first rule, the sum is 100 + 17.5 + 50 + 8.75 = 176.25 (or more simply 150 + 26.25), which decimalised gives 1.7625 as the constant.

Exercise 6

Work out the constant required from the data given:

a. VAT 12%, mark-up 55%
b. VAT 13%, mark-up 45%
c. VAT 14%, mark-up 60%
d. VAT 16%, mark-up 65%
e. VAT 16.5%, mark-up 75%
f. VAT 18%, mark-up 70%
g. VAT 21%, mark-up 60%
h. VAT 17.5%, mark-up 55%

Because VAT has to be collected at each stage in the life of a product or service, Customs and Excise issue a form to be filled in quoting the total sales and the VAT to be paid. This does involve a calculation, but fortunately for the payee the maths has already been done. The form quotes a number by which the payee divides their total sales income in order to reach the VAT payable to Customs and Excise.

It is useful to understand how this figure is calculated, because the same procedure can be used to calculate the original figure after a percentage increase has been added. For instance, it is often stated in the media that a person has been given a certain percentage increase in salary, without the news item divulging the original salary. The relevant equation is calculated to a base of 100% plus the percentage increase. This new total is then divided by the percentage increase, giving a figure that is divided into the total, the result being the percentage increase.

Let's look at a simple increase of 10%. The sum

is 100 + 10 = 110; this is then divided by 10, which gives 11; the total sum is therefore 100 + 10 = 110 ÷ 10 = 11: the VAT payable is the total takings divided by 11.

The current VAT rate is 17.5%, so using this figure the sum is 100 + 17.5 ÷ 17.5 = 6.714, this is the dividing figure used to calculate the VAT content. To give another example, salary plus increase of 25% is £125,000, what was the original salary? Following the procedure I have outlined, 100% + 25% = 125 ÷ 25 = 5; £125,000 ÷ 5 = £25,000; £125,000 – £25,000 = £100,000; therefore the original salary was £100,000.

Exercise 7

Calculate the following:

a. Total sales £38,659, VAT rate 14%, what is the VAT payable?
b. Total sales £69,387, VAT rate 16%, what is the VAT payable?
c. Total sales £173,800, VAT rate 15%, what is the VAT payable?
d. A person receives a 27% increase in salary, making his new salary £136,000, what was his original salary?
e. A person receives a 43% increase in salary, making his new salary £295,000, what was his original salary?

Negative Equity

Much has been written about negative equity in recent years, how people have found themselves in this situation without realising that it could happen. Simply, it was related to the equity (value) of their house and their ability to continue paying their mortgage.

Negative equity became a widespread problem because people bought houses in an economic climate of low inflation that was causing house prices to rise rapidly. At the same time the lenders (banks and building societies) had excessive amounts of deposit money. In order to earn money with this surplus, they were allowing people to take out loans of up to 100% of the house value, and at many more times a person's earning rate than was sound practice. When a downturn came in the economy, money became tighter and people lost their jobs or their earnings were reduced, which led to situations where they could not keep up their payments to the lender.

In this type of situation owners are forced into having to sell their houses. More houses on the market coupled with less money around means that house prices fall. When the lenders want their money, people cannot realise the value in their house. Their house is now worth less than they owe and they are thus in negative equity. But they still owe the lender, and so ultimately they are bankrupt.

At the time of writing, the country is in a period of very low interest rates and very high house prices. Again, this is a recipe for negative equity. The problem is that when interest rates are low, small rate rises can mean more than when rates are high. For example, a person has a monthly repayment amount of £500 at a mortgage rate of 12.5%. If the rate is increased by 2.5% to 15%, this is a rise of one fifth or a 20% increase, which in money terms means £100; but if this £500 monthly repayment was at a rate of 5%, and the rate was increased by 2.5%, this is an increase of 50%: in money terms this means an increase of £250 a month, making a total of £750 a month. Beware of borrowing a lot of money just because interest rates are low. There is a nasty shock when they rise.

The Euro

Whilst Britain is still not in the euro, the euro is used in all business transactions conducted on the Continent of Europe and by all travellers and holidaymakers who wish to go to Europe. It is quoted in two different ways, either by the value of one pound sterling to its equivalent in euros, or by the pence value of one euro. Both these methods are used daily by the banks, and by exchange bureaux.

To give an example in figures, on a given day one euro can be quoted as being worth 70 pence, or one pound sterling can be quoted as being worth 1.428 euros.

Both the above figures are comparable; if 100 (the number of pence in one pound) is divided by 70, the result is 1.428 euros to the pound.

The following examples of exchange transactions do not take into consideration any charges levied by the office making the transaction.

How many euros will your sterling buy if the rate is, one euro equals 70 pence. This can be calculated two ways: either dividing by .7 (70 pence is .7 of one pound) or multiplying by 1.428. Of course this is only using the above figures; the correct conversion figures for that day have to be found when making any calculation. But using the above figure:

£500 = 714 euros
£750 = 1,071 euros
£200 = 285 euros

How much are the euros you have left worth in sterling? Again, using the above figures, either multiply by .7 or divide by 1.428; as you can see, it is exactly the opposite calculation.

500 euros = £350
864 euros = £605
372 euros = £260

5

ANSWERS

Exercise 1

a. .001 = 1/1000 = one thousandth
b. 10 = ten
c. 1,000 = one thousand
d. .1 = 1/10 = one tenth
e. 10,000 = ten thousand
f. .01 = 1/100 = one hundredth
g. 100 = one hundred

Exercise 2

a. .125
b. .333
c. .111
d. .2
e. .142
f. .0666
g. .05
h. .0285

j. .025
k. .02
l. .625
m. .833
n. .571
p. .777
r. .375

Exercise 3

a. £662.63
b. £393.39
c. £1,289.87
d. £3,797.31
e. £2,870.73

Exercise 4

a. .13
b. .42
c. .095
d. .0175
e. 1.22
f. 2.03
g. .0875
h. .005

Exercise 5

a. £585.41

b. £1,278.68
c. £1,366.18
d. £17,389.11
e. £16,594.47
f. £10,441.43

Exercise 6

a. 173.6
b. 163.85
c. 182.4
d. 191.4
e. 203.875
f. 200.6
g. 193.6
h. 182.125

Exercise 7

a. £4,747.63
b. £9,570.62
c. £22,671.53
d. £107,082.30
e. £206,278.20

6

APPENDIX

Formulae

Decimal Equivalent of a Fraction
To find the decimal equivalent of a fraction, divide the numerator (the number above the line) by the denominator (the number below the line).

Simple Interest
APR is the annual percentage rate, the amount added to the capital at the end of one year, when a sum of money has been on deposit for one year. If a sum of money is on deposit for a shorter term than one year, the annual rate will be used pro rata, (for the length of that term). For instance, a sum on deposit for six months will yield half the annual amount of interest.

Compound Interest
The amount of interest added to the capital plus interest when a sum of money is left on deposit longer than the term specified. An annual term would

mean that it would begin to compound after the end of one year.

Interest Rates other than Annual
A monthly rate of interest is the amount calculated each month on the capital plus the monthly addition of interest. In one year this would mean twelve calculations.

Multiplication or Division by a Positive Decimal
When a sum is multiplied by a positive decimal figure (for instance greater than one, e.g.1.7), it will increase the size of the sum e.g.$100 \times 1.7 = 170$: when a sum is divided by a positive decimal, it will decrease the size of the sum ($100 \div 1.7 = 58.82$).

Multiplication or Division by a Negative Decimal
When a sum is multiplied by a negative decimal figure (less than one, e.g. 0.7) it will reduce the size of the sum, e.g. $100 \times 0.7 = 70$: when a sum is divided by a negative decimal it will increase the size of the sum, $100 \div 0.7 = 142.85$.

Calculating Interest Rates by Positive Multiplication
First turn the interest rate into a decimal. $10\% = .1$, add $1 = 1.1$, then multiply the capital sum by 1.1 (the interest is added to the capital). If the capital sum is multiplied by .1 the interest only will be shown.

Finding the Percentage Increase on a Known Amount
First calculate the 'increase' by deducting the original amount from the increased amount. Then divide the

46

increase by the original and multiply the result by a hundred.

$$\frac{\text{Increase}}{\text{Original amount}} \times \frac{100}{1} = \% \text{ increase}$$

Finding the Percentage Decrease on a Known Amount
First determine the original amount and the decreased amount. Divide the original amount by the decreased amount, then divide 100 by this figure. This will give the percentage of the decrease.
  Original amount 1,000, decreased amount 800;
  1,000 − 800 = 200
  1,000 ÷ 200 = 5, 100 ÷ 5 = 20%

Finding the Percentage Increase Needed to Make Good a Loss from a Capital Amount
Divide the amount lost by the remaining capital and multiply by 100.
  Original amount 100, capital left 80, loss 20.
  $\frac{20}{80} \times \frac{100}{1} = 25\%$

Finding the Capital Needed to Generate a Required Amount at a Known Interest Rate
Divide the amount required by 100 plus the known interest rate. This equals the capital required. 5500 required, interest rate 10%, the sum is:
  $\frac{5500}{100 + 10} = 5000$

Finding the Constant on a Calculator
The easiest way is to calculate to 10% (so the sum

is 100 + 10 as a decimal equating to 1.1) and then to calculate to 100; so the sum is 1.1 × 100 = or 100 × 1.1 = , the first = sign will show 110, then, pressing the = sign again will either show 121 or some other number, the way that it shows 121 is the constant for that calculator.

Finding the VAT Content of a Sum of Money that Includes VAT
First divide 100 plus the VAT rate, say, 10% (=110) by the VAT rate (110 ÷ 10 = 11). Next divide the sum of money (say 20,000) by 11 (20,000 ÷ 11 is 1,818.20). This is the VAT content. As the rate changes the dividing sum must be recalculated at the first stage.

Euro Conversions
The euro rate can be shown as either the value in pence or as a euro equivalent to the pound sterling. Assuming a euro to be worth 70 pence, the euro equivalent to one pound will be 1.428; this means that one pound sterling will buy 1.428 euros. Converting pounds to euros, multiply by 1.428, or divide by .7. Converting euros to pounds, multiply by .7 or divide by 1.428.

The above calculations are only at the rate shown. The up-to-date rates can be found on CEEFAX or at any bank or exchange bureau.